The Real Logarithm: Theorems and Counterexamples

by

Richard Shedenhelm

To David Cohen

CONTENTS

INTRODUCTION

The real logarithm is to be contrasted with the complex logarithm, that latter accepting complex numbers in its domain.

The first section of this work treats what most textbooks call the "laws of logarithms." Although there are not universally accepted names for such theorems, I have found in teaching that definite descriptions (e.g., "The Power Rule") are helpful to the beginning student.

The second section builds on the first, establishing theorems that are apparently useful in various disciplines.

The last section is especially innovative. Certain errors are commonly committed by the new student. The goal of this section is to convince the reader that certain operations involving logarithms are invalid and undependable. The method I use involves constructing counterexamples that are simple to grasp, employing the theorems proved in earlier sections along with basic claims such as

$$0 \neq 1.$$

As an example of this method applied to algebra, we could provide a counterexample to the erroneous claim that for all real numbers

$$(x + y)^2 = x^2 + y^2.$$

For if we set $x = y = 1$, we get the unhappy result

$$(1 + 1)^2 = 2^2 = 4 \neq 2 = 1 + 1 = 1^2 + 1^2.$$

In this work, the italicized lower-case Roman letters a, b, c shall stand for any real positive numbers other than 1. They shall be used for bases. The letter m shall stand for any real number. It shall be used for exponents. The letters x, y, z shall stand for any positive real numbers. They shall be used as terms in the arguments of the logarithmic functions. Unless otherwise stated, all statements employing such variables shall be taken to hold universally, without exception.

My love for logarithms began in a UCLA precalculus course during the fall of 1987. The "laws of logarithms" present a fine small set of theorems provable in a variety of ways from the definition of a logarithm. I was immediately captivated by the desire to prove the laws in the most direct manner possible from the definition. My first attempts were encouraged by my then teacher, David Cohen. Ever since then, I have been grateful to him for his warm laudations. For this reason, I dedicate this work to him.

The "target audience" for this work is someone like me around age 20: A person lacking in calculus instruction, but passionately curious to see a complete, concise, and clear treatment of the subject of logarithms.

At the time of my first musings in 1987, a very intelligent friend of mine dryly rebuked my proofs as "too simple" to be legitimate. If my proofs herein are "too simple" and yet valid, I happily plead guilty.

Athens, Georgia, October 25, 2015.

Richard Shedenhelm

Basic Logarithmic Properties

Definition 1: $$\log_a(x) = m$$
if and only if

$$a^m = x.$$

In "$\log_a(x)$," "a" is the "base" of the logarithm, and the expression between the parentheses--here denoted by "x"--is called the "argument" of the logarithmic function. The "m" in the first equation above is the "value" of the logarithmic function.

Remark: "A logarithm is an exponent."

Example: $$\log_2(8) = 3,$$
since

$$2^3 = 8;$$

and vice versa.

Theorem 1 (Logarithmic Identity Property):
$$\log_a(a) = 1.$$

Preliminary Remark: In words: The logarithm of any number to the base of the same number is always equal to 1.

Proof: $$a^1 = a.$$
Therefore,

$$\log_a(a) = 1,$$

by Definition 1.

■

Theorem 2 (Logarithm of One): $\log_a(1) = 0.$

Preliminary Remark: In words: The logarithm of 1 is always equal to 0.

Proof: $$a^0 = 1.$$
Therefore,

$$\log_a(1) = 0,$$

by Definition 1.

■

Theorem 3 (Inverse Property of a Logarithm):
$$\log_a(a^m) = m.$$

Proof:
$$a^m = a^m.$$
Therefore,
$$\log_a(a^m) = m,$$
by Definition 1.

∎

Theorem 4 (Inverse Property of an Exponential):
$$a^{\log_a(x)} = x.$$

Proof:
$$\log_a(x) = \log_a(x).$$
Therefore,
$$a^{\log_a(x)} = x,$$
by Definition 1.

∎

Theorem 5 (Product Rule): $\qquad\qquad \log_a(x \cdot y) = \log_a(x) + \log_a(y).$

Preliminary Remark: In words: The logarithm of a product equals the sum of the logarithms of the two factors.

Proof:
$$a^{\log_a(x) + \log_a(y)} = a^{\log_a(x)} \cdot a^{\log_a(y)} = x \cdot y,$$
by Theorem 4. Therefore,
$$\log_a(x \cdot y) = \log_a(x) + \log_a(y),$$
by Definition 1.

∎

Remark: Cf. Counterexamples 1, 2, and 3.

Example: $\quad \log_2(32) = \log_2(8 \cdot 4) = \log_2(8) + \log_2(4) = \log_2(2^3) + \log_2(2^2) = 3 + 2 = 5$

Theorem 6 (Quotient Rule): $$\log_a\left(\tfrac{x}{y}\right) = \log_a(x) - \log_a(y).$$

Preliminary Remark: In words: The logarithm of a quotient equals the logarithm of the numerator minus the log of the denominator.

Proof:
$$a^{\log_a(x) - \log_a(y)} = \frac{a^{\log_a(x)}}{a^{\log_a(y)}} = \frac{x}{y},$$
by Theorem 4. Therefore,

$$\log_a\left(\tfrac{x}{y}\right) = \log_a(x) - \log_a(y),$$

by Definition 1.

■

Remark: Cf. Counterexample 4.

Theorem 7 (Power Rule): $$\log_a(x^m) = m \cdot \log_a(x).$$

Preliminary Remark: In words: The logarithm of a power equals the exponent times the logarithm of the power's base.

Proof:
$$x^m = \left[a^{\log_a(x)}\right]^m,$$
by Theorem 4.

Hence,

$$\left[a^{\log_a(x)}\right]^m = a^{\log_a(x) \cdot m} = a^{m \cdot \log_a(x)}.$$

Therefore,

$$a^{m \cdot \log_a(x)} = x^m.$$

$$\log_a(x^m) = m \cdot \log_a(x),$$

by Definition 1.

■

Remark: Cf. Counterexamples 5 and 6.

Theorem 8 (Change of Base Formula): $\log_a(x) = \frac{\log_b(x)}{\log_b(a)}.$

Proof:
by Theorem 4.

$$\log_b(x) = \log_b\left(a^{\log_a(x)}\right),$$

by Theorem 7. Hence,

$$\log_b\left(a^{\log_a(x)}\right) = \log_a(x) \cdot \log_b(a),$$

Therefore,

$$\log_a(x) \cdot \log_b(a) = \log_b(x).$$

$$\log_a(x) = \frac{\log_b(x)}{\log_b(a)}.$$

$\blacksquare$

Example: $\log_2(x) = \frac{\log_e(x)}{\log_e(2)} = \frac{\log_{10}(x)}{\log_{10}(2)}.$

Some Consequences of the Basic Logarithmic Properties

Theorem 9 (taking logs of equals): If
$$x = y,$$
then
$$\log_a(x) = \log_a(y).$$

Preliminary Remark: In words: Equal quantities have equal logarithms (of the same base).

Proof: Suppose that
$$x = y.$$
Hence,
$$a^{\log_a(x)} = a^{\log_a(y)},$$
by Theorem 4. So,
$$\log_a\left(a^{\log_a(y)}\right) = \log_a(x),$$
by Definition 1. Thus,
$$\log_a(y) = \log_a(x),$$
by Theorem 4. Therefore,
$$\log_a(x) = \log_a(y).$$

■

Remark: This theorem is really an expression that logarithms are *functions*.

Theorem 10 (eliminating logs): If
$$\log_a(x) = \log_a(y),$$
then
$$x = y.$$

Preliminary Remark: In words: Equal logarithms have equal arguments.

Proof: Suppose that
$$\log_a(x) = \log_a(y).$$
Hence,
$$a^{\log_a(y)} = x$$
and
$$a^{\log_a(x)} = y,$$
by Definition 1.
$$a^{\log_a(x)} = a^{\log_a(x)}.$$
So,
$$a^{\log_a(y)} = a^{\log_a(x)},$$
by substitution. Therefore,
$$x = y,$$
by substitution.

■

5

Remark: This theorem is really an expression that logarithms are *one-to-one* functions.

Theorem 11 (Change of Base for Exponentials):
$$a^m = b^{m \cdot \log_b(a)}.$$

Proof:
$$a^m = b^{\log_b(a^m)},$$
by Theorem 4.

$$b^{\log_b(a^m)} = b^{m \cdot \log_b(a)},$$

by Theorem 7. Therefore,

$$a^m = b^{m \cdot \log_b(a)}.$$

■

Example:
$$2^m = e^{m \cdot \log_e(2)} = 10^{m \cdot \log_{10}(2)}.$$

Problem: Express $5^x \cdot \log_9(x)$ in base e.

Solution:
$$5^x = e^{x \cdot \log_e(5)},$$
by Theorem 11.

$$\log_9(x) = \frac{\log_e(x)}{\log_e(9)},$$

by Theorem 8. Therefore,

$$5^x \cdot \log_9(x) = e^{x \cdot \log_e(5)} \cdot \frac{\log_e(x)}{\log_e(9)}.$$

Theorem 12:
$$\log_a\left(\tfrac{x}{y}\right) + \log_a\left(\tfrac{y}{x}\right) = 0.$$

Proof:
$$\log_a\left(\tfrac{x}{y}\right) + \log_a\left(\tfrac{y}{x}\right) = \log_a\left(\tfrac{x}{y} \cdot \tfrac{y}{x}\right),$$
by Theorem 5.

$$\log_a\left(\tfrac{x}{y} \cdot \tfrac{y}{x}\right) = \log_a(1) = 0,$$

by Theorem 2. Therefore,

$$\log_a\left(\tfrac{x}{y}\right) + \log_a\left(\tfrac{y}{x}\right) = 0.$$

■

Theorem 13:

$$\log_a\left(x^{\log_a(x)}\right) = [\log_a(x)]^2.$$

Proof:
by Theorem 7.

$$\log_a\left(x^{\log_a(x)}\right) = \log_a(x) \cdot \log_a(x),$$

$$\log_a(x) \cdot \log_a(x) = [\log_a(x)]^2.$$

Therefore,

$$\log_a\left(x^{\log_a(x)}\right) = [\log_a(x)]^2. \qquad \blacksquare$$

Theorem 14:

$$\log_a\left(x^{\log_a(x^{\log_a(x)})}\right) = [\log_a(x)]^3.$$

Proof:
by Theorem 13.

$$\log_a\left(x^{\log_a(x^{\log_a(x)})}\right) = \log_a\left(x^{[\log_a(x)]^2}\right),$$

by Theorem 7.

$$\log_a\left(x^{[\log_a(x)]^2}\right) = [\log_a(x)]^2 \cdot \log_a(x),$$

$$[\log_a(x)]^2 \cdot \log_a(x) = [\log_a(x)]^3.$$

Therefore,

$$\log_a\left(x^{\log_a(x^{\log_a(x)})}\right) = [\log_a(x)]^3. \qquad \blacksquare$$

Theorem 15:

$$\log_a(x + y) = \log_a(x) + \log_a\left(1 + \tfrac{y}{x}\right).$$

Proof:
But,

$$\log_a(x + y) = \log_a\left(x + x \cdot \left[\tfrac{y}{x}\right]\right) = \log_a\left(x \cdot \left[1 + \tfrac{y}{x}\right]\right).$$

$$\log_a\left(x \cdot \left[1 + \tfrac{y}{x}\right]\right) = \log_a(x) + \log_a\left(1 + \tfrac{y}{x}\right),$$

by Theorem 5. Therefore,

$$\log_a(x + y) = \log_a(x) + \log_a\left(1 + \tfrac{y}{x}\right). \qquad \blacksquare$$

Theorem 16:
$$\log_a(x+y) = \log_a(x) + \log_a\left(1 + a^{\log_a(y) - \log_a(x)}\right).$$

Proof:
$$\log_a(x+y) = \log_a(x) + \log_a\left(1 + \frac{y}{x}\right),$$
by Theorem 15.

$$a^{\log_a(y)} = y$$
and
$$a^{\log_a(x)} = x,$$
by Theorem 4. Hence,

$$\log_a(x) + \log_a\left(1 + \frac{y}{x}\right) = \log_a(x) + \log_a\left(1 + \frac{a^{\log_a(y)}}{a^{\log_a(x)}}\right) =$$
$$= \log_a(x) + \log_a\left(1 + a^{\log_a(y) - \log_a(x)}\right).$$

Therefore,
$$\log_a(x+y) = \log_a(x) + \log_a\left(1 + a^{\log_a(y) - \log_a(x)}\right).$$ ∎

Theorem 17: For $x > y$,
$$\log_a(x-y) = \log_a(x) + \log_a\left(1 - \frac{y}{x}\right).$$

Preliminary Remark: The purpose of the constraint $x > y$ is to prevent $x - y \leq 0$ and $1 - \frac{y}{x} \leq 0$ in the arguments of the logarithmic functions, which, per Definition 1, are forbidden.

Proof: For $x > y$,
$$\log_a(x - y) = \log_a(x + [-y]).$$
But,
$$\log_a(x + [-y]) = \log_a(x) + \log_a\left(1 + \frac{[-y]}{x}\right),$$
by Theorem 15.
$$\log_a(x) + \log_a\left(1 + \frac{[-y]}{x}\right) = \log_a(x) + \log_a\left(1 - \frac{y}{x}\right).$$
Therefore,
$$\log_a(x - y) = \log_a(x + [-y]).$$ ∎

Theorem 18: For $x > y$,

$$\log_a(x - y) = \log_a(x) + \log_a\left(1 - a^{\log_a(y) - \log_a(x)}\right).$$

Proof: For $x > y$,

$$\log_a(x - y) = \log_a(x) + \log_a\left(1 - \frac{y}{x}\right),$$

by Theorem 17.

and

$$a^{\log_a(y)} = y$$

$$a^{\log_a(x)} = x,$$

by Theorem 4. Hence,

$$\log_a(x) + \log_a\left(1 - \frac{y}{x}\right) = \log_a(x) + \log_a\left(1 - \frac{a^{\log_a(y)}}{a^{\log_a(x)}}\right) =$$
$$= \log_a(x) + \log_a\left(1 - a^{\log_a(y) - \log_a(x)}\right).$$

Therefore,

$$\log_a(x - y) = \log_a(x) + \log_a\left(1 - a^{\log_a(y) - \log_a(x)}\right).$$

■

Theorem 19:

$$\log_a(x) \cdot \log_b(a) = \log_b(x).$$

Proof: By Theorem 8.

■

Theorem 20:

$$a^{\log_b(c)} = c^{\log_b(a)}.$$

Proof:
by Theorem 11.

$$a^{\log_b(c)} = c^{\log_b(c) \cdot \log_c(a)},$$

But,

$$c^{\log_b(c) \cdot \log_c(a)} = c^{\log_c(a) \cdot \log_b(c)}.$$

by Theorem 19. Therefore,

$$c^{\log_c(a) \cdot \log_b(c)} = c^{\log_b(a)},$$

$$a^{\log_b(c)} = c^{\log_b(a)}.$$

■

Problem: Evaluate $10^{\log_{100}(9)}$.

Solution:
by Theorem 20.

$$10^{\log_{100}(9)} = 9^{\log_{100}(10)},$$

by Theorem 3.

$$9^{\log_{100}(10)} = 9^{\log_{100}\left(100^{\frac{1}{2}}\right)} = 9^{\frac{1}{2}},$$

Therefore,

$$9^{\frac{1}{2}} = 3.$$

$$10^{\log_{100}(9)} = 3.$$

Theorem 21:

$$\log_a(b) = \frac{1}{\log_b(a)}.$$

Proof:
by Theorem 8.

$$\log_a(b) = \frac{\log_b(b)}{\log_b(a)},$$

$$\log_b(b) = 1,$$

by Theorem 1. Therefore,

$$\log_a(b) = \frac{1}{\log_b(a)}.$$

■

Theorem 22:

$$\log_a(b) \cdot \log_b(a) = 1.$$

Proof: By Theorem 21.

■

Theorem 23:

$$\log_a(x) \cdot \log_b(a) \cdot \log_c(b) = \log_c(x).$$

Proof:
$$\log_a(x) \cdot \log_b(a) \cdot \log_c(b) = [\log_a(x) \cdot \log_b(a)] \cdot \log_c(b) =$$
$$= \log_b(x) \cdot \log_c(b) = \log_c(x),$$
by Theorem 19.

■

Theorem 24:
$$\log_a(x) \cdot \log_b(y) \cdot \log_c(z) = \log_b(x) \cdot \log_c(y) \cdot \log_a(z).$$

Proof:
$$\log_a(x) \cdot \log_b(y) \cdot \log_c(z) = \left[\frac{\log_b(x)}{\log_b(a)}\right] \cdot \left[\frac{\log_c(y)}{\log_c(b)}\right] \cdot \left[\frac{\log_a(z)}{\log_a(c)}\right],$$

by Theorem 8.
$$\left[\frac{\log_b(x)}{\log_b(a)}\right] \cdot \left[\frac{\log_c(y)}{\log_c(b)}\right] \cdot \left[\frac{\log_a(z)}{\log_a(c)}\right] = \frac{\log_b(x) \cdot \log_c(y) \cdot \log_a(z)}{\log_b(a) \cdot \log_c(b) \cdot \log_a(c)} = \frac{\log_b(x) \cdot \log_c(y) \cdot \log_a(z)}{\log_a(a)},$$

by Theorem 23.
$$\frac{\log_b(x) \cdot \log_c(y) \cdot \log_a(z)}{\log_a(a)} = \frac{\log_b(x) \cdot \log_c(y) \cdot \log_a(z)}{1} = \log_b(x) \cdot \log_c(y) \cdot \log_a(z),$$

by Theorem 1. Therefore,
$$\log_a(x) \cdot \log_b(y) \cdot \log_c(z) = \log_b(x) \cdot \log_c(y) \cdot \log_a(z).$$
∎

Theorem 25: For $x \neq 1$,
$$\frac{\log_a(x)}{\log_{a \cdot b}(x)} = 1 + \log_a(b).$$

Preliminary Remark: The purpose of the extra constraint $x \neq 1$ is to prevent $\log_{a \cdot b}(x) = 0$, which would render "$\frac{\log_a(x)}{\log_{a \cdot b}(x)}$" undefined.

Proof:
$$\log_a(x) = \log_a(x)$$
Hence, for $x \neq 1$,
$$\frac{\log_a(x)}{\log_{a \cdot b}(x)} = \log_a(x) \cdot \log_x(a \cdot b),$$

by Theorem 21.
$$\log_a(x) \cdot \log_x(a \cdot b) = \log_a(x) \cdot [\log_x(a) + \log_x(b)],$$

by Theorem 5.
$$\log_a(x) \cdot [\log_x(a) + \log_x(b)] = \log_a(x) \cdot \log_x(a) + \log_a(x) \cdot \log_x(b) =$$
$$= 1 + \log_a(x) \cdot \log_x(b),$$

by Theorem 22.
$$1 + \log_a(x) \cdot \log_x(b) = 1 + \log_x(b) \cdot \log_a(x) = 1 + \log_a(b),$$

by Theorem 19. Therefore,
$$\frac{\log_a(x)}{\log_{a \cdot b}(x)} = 1 + \log_a(b).$$
∎

Theorem 26: For $m \neq 0$,
$$\log_{a^m}(x) = \frac{\log_a(x)}{m}.$$

Proof: For $m \neq 0$,
$$\log_{a^m}(x) = \frac{\log_a(x)}{\log_a(a^m)},$$

by Theorem 8.
$$\frac{\log_a(x)}{\log_a(a^m)} = \frac{\log_a(x)}{m},$$

by Theorem 3. Therefore,
$$\log_{a^m}(x) = \frac{\log_a(x)}{m}.$$

■

Theorem 27: For $m \neq 0$,
$$\log_{a^m}(x^m) = \log_a(x).$$

Preliminary Remark: "Base and argument are like powers."

Proof: For $m \neq 0$,
$$\log_{a^m}(x^m) = m \cdot \log_{a^m}(x),$$

by Theorem 7.
$$m \cdot \log_{a^m}(x) = m \cdot \left[\frac{\log_a(x)}{m}\right],$$

by Theorem 26.
$$m \cdot \left[\frac{\log_a(x)}{m}\right] = \log_a(x).$$

Therefore,
$$\log_{a^m}(x^m) = \log_a(x).$$

■

Example:
$$\log_{1000}(729) = \log_{10^3}(9^3) = \log_{10}(9).$$

Theorem 28: $$-\log_a(x) = \log_a\left(\tfrac{1}{x}\right) = \log_{\frac{1}{a}}(x).$$

Preliminary Remark: "Base and argument are reciprocals."

Proof:
by Theorem 7.

$$-\log_a(x) = -1 \cdot \log_a(x) = \log_a(x^{-1}),$$

$$\log_a(x^{-1}) = \log_a\left(\tfrac{1}{x}\right).$$

Hence,

$$-\log_a(x) = \log_a\left(\tfrac{1}{x}\right).$$

Furthermore,

$$-\log_a(x) = \frac{\log_a(x)}{-1} = \log_{a^{-1}}(x),$$

by Theorem 26.

$$\log_{a^{-1}}(x) = \log_{\frac{1}{a}}(x).$$

So,

$$-\log_a(x) = \log_{\frac{1}{a}}(x).$$

Therefore,

$$-\log_a(x) = \log_a\left(\tfrac{1}{x}\right) = \log_{\frac{1}{a}}(x).$$

$\blacksquare$

Theorem 29: $$\log_a(x) + \log_{\frac{1}{a}}(x) = 0.$$

Proof:
by Theorem 28.

$$\log_a(x) + \log_{\frac{1}{a}}(x) = \log_a(x) + \log_a\left(\tfrac{1}{x}\right),$$

$$\log_a(x) + \log_a\left(\tfrac{1}{x}\right) = \log_a\left(\tfrac{x}{1}\right) + \log_a\left(\tfrac{1}{x}\right).$$

Furthermore,

$$\log_a\left(\tfrac{x}{1}\right) + \log_a\left(\tfrac{1}{x}\right) = 0,$$

by Theorem 12. Therefore,

$$\log_a(x) + \log_{\frac{1}{a}}(x) = 0.$$

$\blacksquare$

Theorem 30:

$$x^{\log_a[\log_a(x)]\cdot[\log_a(x)]^{-1}} = \log_a(x).$$

Proof:

$$\log_a(x) = \log_a(x).$$

Hence,

$$\log_a(x) \cdot [\log_a(x)]^{-1} = 1.$$

Furthermore,

$$\log_a[\log_a(x)] \cdot 1 = \log_a[\log_a(x)].$$

So,

$$\log_a[\log_a(x)] \cdot \{\log_a(x) \cdot [\log_a(x)]^{-1}\} = \log_a[\log_a(x)],$$

by substitution. Thus,

$$\{\log_a[\log_a(x)] \cdot [\log_a(x)]^{-1}\} \cdot \log_a(x) = \log_a[\log_a(x)],$$

Hence,

$$\log_a\big[x^{\log_a[\log_a(x)]\cdot[\log_a(x)]^{-1}}\big] = \log_a[\log_a(x)],$$

by Theorem 7. Therefore,

$$x^{\log_a[\log_a(x)]\cdot[\log_a(x)]^{-1}} = \log_a(x),$$

by Theorem 10.

■

Counterexamples to Some Common Logarithmic Errors

Counterexample 1: For some a, x, and y,
$$\log_a(x + y) \neq \log_a(x) + \log_a(y).$$

Proof: Suppose for the sake of contradiction that
$$\log_a(x + y) = \log_a(x) + \log_a(y).$$
Hence, with $a = 2$ and $x = y = 1$,
$$\log_2(1 + 1) = \log_2(2) = \log_2(1) + \log_2(1),$$
by substitution. But
$$\log_2(2) = 1 \text{ and } \log_2(1) = 0,$$
by Theorems 1 and 2. So,
$$1 = 0 + 0 = 0.$$
Contradiction. ∎

Counterexample 2: For some a, x, and y,
$$\log_a(x \cdot y) \neq \log_a(x) \cdot \log_a y$$

Proof: Suppose for the sake of contradiction that
$$\log_a(x \cdot y) = \log_a(x) \cdot \log_a(y).$$
Hence, with $a = x = 2$ and $y = 1$,
$$\log_2(2 \cdot 1) = \log_2(2) = \log_2(2) \cdot \log_2(1),$$
by substitution. But
$$\log_2(2) = 1 \text{ and } \log_2(1) = 0,$$
by Theorems 1 and 2. So,
$$1 = 1 \cdot 0 = 0.$$
Contradiction. ∎

Counterexample 3: For some a, x, and y,
$$\log_a(x + y) \neq \log_a(x) \cdot \log_a(y).$$

Proof: Suppose for the sake of contradiction that
$$\log_a(x + y) = \log_a(x) \cdot \log_a(y).$$
Hence, with $a = 2$ and $x = y = 1$,
$$\log_2(1 + 1) = \log_2(2) = \log_2(1) \cdot \log_2(1),$$
by substitution. But
$$\log_2(2) = 1 \text{ and } \log_2(1) = 0,$$
by Theorems 1 and 2. So,
$$1 = 0 \cdot 0 = 0.$$
Contradiction. ∎

15

Counterexample 4: For some a, x, and y,
$$\frac{\log_a(x)}{\log_a(y)} \neq \log_a(x - y)$$

Proof: Suppose for the sake of contradiction that
$$\frac{\log_a(x)}{\log_a(y)} = \log_a(x - y)$$
Hence, with $a = y = 2$ and $x = 4$,
$$\frac{\log_2(4)}{\log_2(2)} = \log_2(4 - 2),$$
by substitution. But
$$\log_2(4) = \log_2(2^2) = 2 \cdot \log_2(2) = 2 \cdot 1 = 2,$$
By Theorems 7 and 1. Furthermore,
$$\log_2(4 - 2) = \log_2(2) = 1,$$
by Theorem 1. So,
$$\frac{2}{1} = 2 = 1.$$
Contradiction.

Counterexample 5: For some a, m, and x,
$$[\log_a(x)]^m \neq m \cdot \log_a(x).$$

Proof: Suppose for the sake of contradiction that
$$[\log_a(x)]^m = m \cdot \log_a(x).$$
Hence, with $a = m = x = 2$,
$$[\log_2(2)]^2 = 2 \cdot \log_2(2).$$
by substitution. But
$$\log_2(2) = 1,$$
by Theorem 1. So,
$$1^2 = 1 = 2 \cdot 1 = 2.$$
Contradiction.

Counterexample 6: For some a, m, x, and y,
$$\log_a(x^m + y) \neq m \cdot \log_a(x + y).$$

Proof: Suppose for the sake of contradiction that
$$\log_a(x^m + y) = m \cdot \log_a(x + y).$$
Hence, with $a = x = 2$, $y = 1$, and $m = 0$,
$$\log_2(2^0 + 1) = \log_2(1 + 1) = \log_2(2) = 0 \cdot \log_2(2 + 1) = 0,$$
by substitution. But
$$\log_2(2) = 1,$$
by Theorem 1. So,
$$1 = 0.$$
Contradiction.

■

Counterexample 7: For some a, x, and y,
$$\frac{\log_a(x \cdot y)}{x} \neq \log_a(y).$$

Proof: Suppose for the sake of contradiction that
$$\frac{\log_a(x \cdot y)}{x} = \log_a(y)$$

Hence, with $a = y = 2$ and $x = \frac{1}{2}$,
$$\frac{\log_2\left(\frac{1}{2} \cdot 2\right)}{\frac{1}{2}} = \frac{\log_2(1)}{\frac{1}{2}} = \log_2(2),$$
by substitution. But
$$\log_2(1) = 0 \text{ and } \log_2(2) = 1,$$
by Theorems 2 and 1. So,
$$\frac{0}{\left(\frac{1}{2}\right)} = 0 = 1.$$
Contradiction.

■

"Only he who never plays, never loses"